Developing
Life Skills through
Math &
Science
Games

Developing Life Skills through
Math &
Science
Games

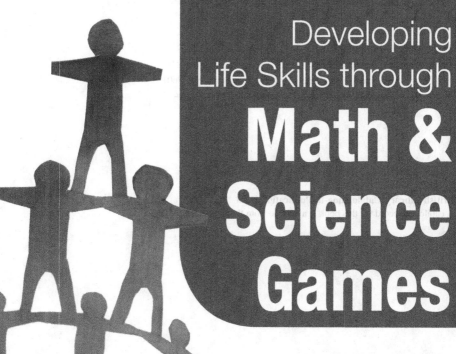

Wee Khee Seah
PhD (NUS), MEdMgt (UniMelb)

Li Yang Ng

Ying Zhen Ang

Reico Ng

Edited by

Beng Lee Lim

World Scientific

NEW JERSEY · LONDON · SINGAPORE · BEIJING · SHANGHAI · HONG KONG · TAIPEI · CHENNAI

Published by

World Scientific Publishing Co. Pte. Ltd.

5 Toh Tuck Link, Singapore 596224

USA office: 27 Warren Street, Suite 401-402, Hackensack, NJ 07601

UK office: 57 Shelton Street, Covent Garden, London WC2H 9HE

British Library Cataloguing-in-Publication Data
A catalogue record for this book is available from the British Library.

ISBN 978-981-4439-81-7 (pbk)

Printed in Singapore.

Foreword

Today's globalised economy presents new challenges. Equipping students with life skills such as leadership and teamwork has become essential to preparing them for success in the modern workplace. Of equal importance are the desire to excel and sound values to guide one's actions.

To meet these demands, the education system has evolved into one that aims to provide students with a holistic education. Schools must also collaborate with stakeholders to nurture responsible citizens who can serve the society with empathy. This book bears testament to the passion and innovation of those who worked on it. It also embodies Hwa Chong Institution's desired outcomes for our students to "Live with Passion and Lead with Compassion". I believe that this book will provide schools, teachers and parents with tools to develop students cognitively, physically, and socially, while encouraging them to be confident, self-directed learners.

I extend my heartiest congratulations to the authors on seeing their work bear fruit. I would also urge you to apply the lessons here and pursue excellence in every endeavour.

Dr Hon Chiew Weng
Principal
HWA CHONG INSTITUTION

Preface

This book is not just another 50 games! This collection of activities incorporates character building and the development of leadership skills using Math and Science concepts. The authors of this book firmly believe that learning life skills through putting ourselves through the suggested activities helps in better understanding and the assimilation of character values in our daily lives. Here within the pages, you will find such an interesting 'marriage' between hard and soft skills.

The versatility of this book allows you to use the games whether you are teaching Math, Science or imparting Life Skills. Just as the world is not one dimensional, you can use any of the principles and concepts to teach the Math and Science curriculum and to reinforce the soft skills both in the classroom and outside of it. In this book, we have mindfully combined both character building and leadership skills into the 50 games, as we believe that they are important for the development of the 'total' individual. These skills, which may be observed in the playing of the games, can be found under the '**Key Aptitudes**' section which indicates Emotional, Social and Intelligence Quotients. Using this format, we hope to nurture each individual holistically by addressing the social, emotional, intelligence realms required for each activity. The reflective questions (*Practical Application*) at the end of each game, serves to help the participants of the games to better appreciate the values of self-awareness and self-development in recognising that life skills and the academics need not be viewed as separate entities.

While each game states detailed step-by-step instructions, you need not follow each game play religiously. These games and their diagrams are here to provide you with ideas and they can be modified to suit your own purpose/s.

Whether one is picking up soft or hard skills, we want to make learning enjoyable. So just as we spent many joyful moments penning this book and testing the games' feasibility, we sincerely hope that you will have sweet memories of your own, learning through play.

Let's do something different for a change!

Let's seriously have some fun!

Acknowledgements

We would like to thank our families and friends who supported us in one way or another.

We are also particularly grateful to those 'guinea pigs' who had spent much time playing these games to check their feasibilities.

CONTENTS

Figure 1. Aptitude Matrix

		ADAPTABILITY	AGILITY	ALERTNESS	APPLICATION	ATTENTIVENESS	COMMUNICATION	CO-OPERATION	CO-ORDINATION	CREATIVITY	DETERMINATION	DEXTERITY	EMPATHY
Biology	Game 6: Body Language			✓		✓	✓						
	Game 13: Dichotomous Frenzy									✓	✓		
	Game 17: Blow Wind Blow			✓		✓							
	Game 21: Who's Walking Now?						✓						
	Game 25: Circle of Trust						✓	✓	✓				
	Game 27: Knotted DNA							✓				✓	
	Game 29: Seeds and Sticks						✓						✓
	Game 31: Living Dominoes				✓		✓						
	Game 35: Virus Attack							✓			✓		
	Game 39: Categorize Me!									✓		✓	
	Game 40: Rhythm of Life												
	Game 43: Mystery In The Food Web						✓						
	Game 49: Limps In Motion							✓				✓	
Chemistry	Game 5: Bouncing Eggs			✓			✓		✓			✓	
	Game 8: Chemical Reaction!						✓	✓					
	Game 12: Under Pressure						✓						
	Game 23: Bumper Particles	✓											
	Game 32: Chemical Creation	✓				✓	✓						
	Game 33: Blast-Off!					✓	✓	✓					
	Game 38: That Sinking Feeling									✓			
	Game 42: Freezing Points						✓	✓					
	Game 45: Breathe!												
	Game 48: Chemi-Who?												
	Game 50: Melting Pot						✓			✓			

| FLEXIBILITY | GOAL SETTING | HONESTY | INNOVATION | INTUITION | INTERPRETATION | MEMORY RECALL | MOTIVATION | MULTI-TASKING | PATIENCE | PERCEPTIBILITY | PERSISTENCE | PROBLEM-SOLVING | QUICK THINKING | RELATIONSHIP MANAGEMENT | RESPECT | RESPONSIVENESS | SELF-CONFIDENCE | SENSITIVITY | SOCIAL AWARENESS | SPONTANEITY | TEAMWORK | THOROUGHNESS | TOLERANCE | TRUST |

Figure 1. (Continued)

		ADAPTABILITY	AGILITY	ALERTNESS	APPLICATION	ATTENTIVENESS	COMMUNICATION	CO-OPERATION	CO-ORDINATION	CREATIVITY	DETERMINATION	DEXTERITY	EMPATHY
Mathematics	Game 2: Mental Arithmetic								✓				
	Game 3: Thinking on Your Feet!								✓				
	Game 10: Lucky 7s							✓	✓		✓		
	Game 14: Lost In Equation						✓					✓	
	Game 15: Magic 21												
	Game 18: Splish, Splash, Splosh		✓				✓						
	Game 20: Get Into Shape						✓					✓	
	Game 37: One Blind Mouse						✓						
	Game 47: Same Train						✓	✓	✓				
Physics	Game 1: Back-Ground Music	✓					✓						
	Game 4: Port, Starboard and Core							✓			✓		
	Game 7: Stick Together	✓						✓				✓	
	Game 9: Topsy-Turvy							✓				✓	
	Game 11: Bombs-Away!!						✓	✓				✓	
	Game 16: Tumble Dry								✓				
	Game 19: Reflective Lights		✓				✓	✓					
	Game 22: Long and Short						✓	✓					
	Game 24: Bouncing Balls						✓	✓			✓	✓	
	Game 26: 360° Water			✓							✓		
	Game 28: Cosmos!						✓	✓				✓	
	Game 30: Water Waves						✓	✓				✓	
	Game 34: Losing My Marbles						✓	✓				✓	
	Game 36: Let's Jam!						✓						
	Game 41: Mini-TV!	✓					✓						
	Game 44: Static!	✓											
	Game 46: Flipside							✓	✓				

FLEXIBILITY

GOAL SETTING

HONESTY

INNOVATION

INTUITION

INTERPRETATION

MEMORY RECALL

MOTIVATION

MULTI-TASKING

PATIENCE

PERCEPTIBILITY

PERSISTENCE

PROBLEM-SOLVING

QUICK THINKING

RELATIONSHIP MANAGEMENT

RESPECT

RESPONSIVENESS

SELF-CONFIDENCE

SENSITIVITY

SOCIAL AWARENESS

SPONTANEITY

TEAMWORK

THOROUGHNESS

TOLERANCE

TRUST

About the Authors/Editor

The Authors

Dr Seah Wee Khee graduated with a PhD (NUS) in Life Sciences and has a Masters of Educational Management (MEdMgt) (Top Honours) from the University of Melbourne, Australia. She was also the recipient of the prestigious President Graduate Fellowship in 2004. In 2005, she was awarded the Best Graduate Researcher Award from the Department of Biological Sciences. As an educator, she has a wealth of experience in coaching teachers and graduate students from NUS in research methodology and concepts. Recognising the need for developing well-rounded individuals, she is currently focusing her time on motivating and nurturing individuals holistically into accomplished winners. To date, she has also published 2 books (*50 Math and Science Games for Leadership* and *cheMagic: 50 Chemistry Classics and Magical Tricks*) which are sold internationally, and has several research papers in internationally reviewed journals. She was previously the Head of Research, Innovation and Enterprise in NUS High School of Math and Science.

Ng Li Yang is currently a student in Hwa Chong Institution (College). She always puts in her best in her academic studies, scoring straight As in all her subjects. With a strong interest Mathematics and Science, she has obtained great achievement in several Mathematics and Science competitions like the Singapore Mathematical Competition and the Innovation Programme. Her passion for Chinese traditions and culture earned her a scholarship for the 'Bicultural Studies Programme' (Chinese) presented by the Ministry of Education. As an avid photographer, she enjoys taking pictures from a different angle and to see things from a different perspective. Her photography skills have gained her recognition in competitions like ProjectID organised by NUS. Believing in contributing back to the society and putting a smile on others, she makes an effort to help out at the NUS library and is involved in fundraising activities at organisations such as ARCES (Animal Concerns Research and Education Society).

Ang Ying Zhen is currently an undergraduate studying psychology in a US college. Her focuses are in the fields of behavioural economics and micro-expressions. She used to be the President of the NUS High Student Council. For the last six years, Ying Zhen had been an active debater and participant of the Model United Nations. These experiences had helped her to organise and facilitate numerous camps for students of varying ages and profiles, ranging from nursery to junior college students. The contents of the camps ranged from development in Public speaking, Leadership and team building, or simply having fun.

Ng Wenbin Reico Maynard is currently a student at the National University of Singapore. With a strong passion for Mathematics and Science, Reico has done numerous research projects in the academic field. While in NUS High School of Mathematics and Science (NUSHS), he has attained a gold award in the Singapore Science and Engineering Fair (SSEF) for his research on colloidal crystals. Apart from academics, Reico has taken up leadership roles in numerous clubs and activities. In 2006, he was the head of Internal Affairs of the 2nd NUSHS Student Council, and a year after took up the role of President. As a trainer for NASCANS since 2008, Reico has inspired numerous students through the many workshops organised. He has taught students on subjects ranging from academia to leadership qualities, and often receives positive responses from students on what they have learnt.

The Editor

Mrs Lee Hong Leong (Mdm Lim Beng Lee) is an experienced teacher who has taught in both Secondary and Primary schools for the last 38 years. Her academic expertise in grooming and developing students has been well recognised through her efforts as the Head of Department for English for more than 10 years in both the Primary and Secondary schools. Passionate about education, she also possesses a strong belief and conviction in nurturing soft skills in students and teachers. Her fervour and efforts for character building can be seen in the improvement and advancement of the Student Development Programme in Pei Hwa Presbyterian Primary School where she served as the Head of Student Development for the 5 years. As a respected mentor to students and teachers, she is dedicated to train teachers in soft skills by equipping them with essential skills first and motivating them to become passionate towards developing students holistically. Her efforts in cultivating a strong character programme in the school have been recognised through receiving the prestigious National Day Commendation Award in 2006.

Game 1: Back-Ground Music

1. Set up the room as follows and have players stand behind one of the lines, backfaced to the speakers.

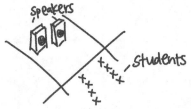

2. Play some music from the speakers. After the music stops, players walk backwards trying to get as close to 3 feet from the speakers without being within.

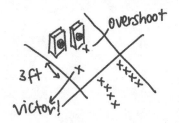

3. Distance of speakers to line can be modified at each turn.

Game 1: Back-Ground Music

Key Aptitude

Emotional Quotient – Adaptability, Intuition
Social Quotient – Communication

Math/Science Concept Applied

Physics – Sound Intensity

Equipment/Logistics

Speaker, Chairs, White Tape, Measuring Tape

Time Required

10-15 minutes

Game Objective

The group with the most people that stand closest to speaker without coming within three feet of it wins.

Group Size

10-16 (Preferably, even numbers)

Procedure

1. A speaker will first be placed on a chair
2. White tape will then be placed around 10 feet away from the speaker, placed perpendicular to the direction the speaker is facing
3. The facilitator will first ask the participants to stand 3 feet away from the speaker while playing a tone/sound on the speaker.

4. The participants are then divided into 2 teams, which will then line up the behind the white tape
5. The facilitator will begin the game by playing another tone/sound
6. Using the impression of the sound played before, a participant from each team will then walk backwards along the white line, aiming to reach as close as possible to reach within three feet
7. The participant at no time is allowed to look at the ground for directions. In addition, no directional guidance should be given to the participant by anyone
8. The player can choose to stop walking backwards at any point, signaled by the raising of his/her hand
9. Once both participants have stopped, a winner will be chosen according to the following conditions:
   - If one participant is standing within three feet of the speakers, the other participant wins by default.
   - If both participants come within three feet of the speakers, it will be considered a draw
   - If both participants did not come within three feet of the speakers, the participant closer to the speaker wins

Possible Variations

For an increase in difficulty, a song could replace the tone played. The participants could also be asked to follow a white line placed in a zig-zag manner while walking backwards

Practical Application

- How did it feel to complete a task based on sound alone?
- What other types of communication could have helped you complete this task more efficiently?

Game 2: Mental Arithmetic

1. Group members are given numbers. Number ①
holds on to the ball.

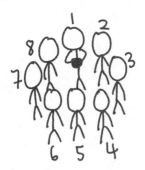

2. Facilitator says a series of numbers, and the
ball is passed in that order.

1,3,7,6

Game 2: Mental Arithmetic

Key Aptitude
Emotional Quotient – Co-ordination, Goal Setting
Social Quotient – Teamwork

Math/Science Concepts Applied
Mathematics – Mental Sums

Equipment/Logistics
Small ball

Time Required
15 minutes

Game Objective
To be the fastest group to throw the ball (in a specified sequence) without it dropping

Group Size
4-10

Procedure

1. The group will first form a circle
2. A player will be a given a ball
3. The player holding the ball shall be known as player Number 1, with the persons to his/her left known as Number 2, 3, 4 and so forth until all the players have a number each

4. The facilitator will then say a random sequence of numbers that will be given to the group. e.g. 6, 3, 4, 8
5. Players will need to pass the ball according to the facilitator's sequence
6. After 2 sequences, the facilitator will then change the players' numbers by either saying a formula (such as asking everyone to multiply their current number by 2) or by shifting the players around
7. The winner will be the group that is able to complete the most number of sequences within a given set time

Possible Variations

Instead of labeling everyone with ascending numbers in a clockwise manner, one can randomly label members with the same set of numbers

Chemistry/ Biology: Other terms can also be used to substitute numbers, such as chemical elements or animal groups

Practical Application

- What were the challenges faced when the facilitator changed a certain aspect of the game?
- How did you adapt to these sudden changes during the game?
- In what way did your fellow group members help you?

Game 3: Thinking on Your Feet!

1. Divide players into 2 groups and have them pair up within the groups.

Group 1   Group 2

2. Each pair is given a shuttlecock. They have to keep it in the air for as long as possible and recite a math sequence at the same time.

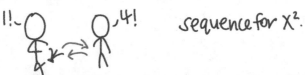

1!~   ,4!   sequence for x^2.

3. Scores are tabulated by adding number of kicks for all pairs in the group!

pair 1: 5	pair 1: 3
pair 2: 15	pair 2: 21
pair 3: 7	pair 3: 8
total: 27	total: 32

Group 2 wins!

Game 3: Thinking on Your Feet!

Key Aptitude

Emotional Quotient – Co-ordination, Multi-tasking
Intelligence Quotient – Memory Recall
Social Quotient – Teamwork

Math/Science Concepts Applied

Mathematics – Sequences and Series, Mental Sums

Equipment/Logistics

Jianzi*

Time Required

10 minutes

Game Objective

To keep the Jianzi in the air as long as possible

Group Size

8-24

Procedure

1. Divide the group into two smaller groups, which are then further divided into pairs
2. The pairs of one group will be handed one Jianzi

---

*Jianzi – an Asian game item which looks like a heavily weighted shuttlecock

3. At the count of 3, the pairs in the 1st group have to start to keep the Jianzi in the air using their one of their feet while shouting out a designated mathematical sequence (such as the 'Fibonnaci' sequence or x^2)
4. The pairs in the 2nd group will have to note the number of kicks made by the pairs in the 1st group while verifying what they shout out (Math Sequence) is correct
5. This will continue until the Jianzi touches the ground or the number shouted by the pair is incorrect
6. The total score for the group is tabulated by adding up the number of kicks scored by all the pairs in the group
7. The group with the higher score wins

Possible Variations

If a Jianzi cannot be found, a small ball can be substituted

Volleyballs may also be used, with groups using their arms to keep the ball in the air instead of their feet

Practical Application

- How did it feel to do multiple tasks concurrently?
- What were some real-life situations where you had to do many things at the same time?
- How did you help your partner in such a situation?

Game 4: Port, Starboard and Core

1. Players are divided into 2 groups.

2. One group stands in a circle, with all members holding a tennis ball each. A member of the other group can give 3 possible commands:

a) Port: Throw the ball in the air and move to the left, catching the next ball.

b) Starboard: Throw the ball in the air and move to the right, catching the next ball.

Game 4: Port, Starboard and Core

Key Aptitude

Emotional Quotient – Flexibility, Determination, Motivation
Social Quotient - Teamwork, Co-operation

Math/Science Concepts Applied

Physics – Time, Force, Gravity

Equipment/Logistics

Tennis Balls

Time Required

10 minutes

Game Objective

To coordinate the actions in the shortest time possible

Group Size

Minimum of 10

Procedure

1. Divide the group into two. Group 1 will play first
2. The second group will elect a facilitator who will give commands to the members of the first group
3. Group 1 will stand in a circle
4. Every player in the Group 1 holds a tennis ball in his right hand
5. The facilitator will shout a command either "Port", "Starboard" or "Core"

6. When he shouts **Starboard**, everyone will toss his tennis balls in the air and moves to the right and catches the ball from the person on the right. When **Port** is heard, everyone moves to the left and catches the ball on the left. When **Core** is shouted, he tosses the ball into the air, claps 3 times and before the ball drops to the floor, catches it
7. Once a ball drops to the floor, the player is disqualified
8. The group that lasts the longest with at least 2 people remaining wins the game

Possible Variations

Extra 'actions' can be added to increase the difficulty, such as when the "Core" command is shouted. In addition to previous actions, one has to spin one round before catching the ball

One can also use small balloons to decrease the difficulty

Practical Application

- How did you feel if there are always some of the members that always missed the tennis balls? How did/could you help these members?
- What problems did you face yourself and how did you overcome them?

Game 5: Bouncing Eggs

1. Facilitator prepares eggs beforehand by soaking them in vinegar for 2 days before the game.

2. Players stand in a circle and bounce the egg to each other without breaking it.

Game 5: Bouncing Eggs

Key Aptitude

Emotional Quotient – Alertness, Dexterity
Social Quotient – Communication, Co-ordination

Math/Science Concept Applied

Chemistry – Acid-Carbonate Reaction

Equipment/Logistics

Vinegar, Eggs

Time Required

10-20 minutes

Game Objective

Bounce eggs from one to another without breaking them

Group Size

10-20

Procedure

1. Facilitator will prepare the eggs by soaking them in vinegar for 2 days before the game
2. Players will stand in a circle
3. Once the game starts, the facilitator will pass one egg to one of the players

4. The aim of the game is for the players to bounce the egg from one another without breaking it
5. To increase the difficulty of the game, the facilitator can add more eggs one at a time to the circle of players

Practical Application

- What principle did you learn from doing this activity?
- How can this game be improved to make it more interesting?

Game 6: Body Language

1. Everyone sits in a circle on chairs, except for 1 person who stands in the middle (called the "scientist"). They will each give themselves a biological name and tell everyone what it is.

2. The scientist will call out these names, and those who are called will stand behind him in the middle of the circle.

"Lung!"

3. At any time, the scientist can either shout "split it" or "live it".

Game 6: Body Language

Key Aptitude

Emotional Quotient - Alertness, Attentiveness
Social Quotient - Communication

Math/Science Concept Applied

Biology - Body Parts

Equipment/Logistics

Chairs

Time Required

20 minutes

Game Objective

To find a seat the soonest possible

Group Size

20-30

Procedure

1. Everyone in the group will sit in a circle on the chairs except for the "scientist" of this game who will stand in the middle of the circle
2. Everyone including the scientist will give himself a biological name that could range from cells like red blood cells or to body parts like the lungs
3. Everyone will tell the others his name and everyone has to try to remember as many names as possible

4. The scientist will call out names and once called, the person with the name will stand behind the scientist
5. This continues until the scientist shout either "split it" or "live it"
6. 'Split it" means that everyone who is standing will try to get a seat while 'live it' means that everyone (including the people sitting) changed their position
7. The last one left standing shall be the scientist for the next round
8. The person who has been the 'scientist' the most number of times after 20 minutes will have to do a forfeit

Practical Application

- How did you adapt to the sudden changes during the game?
- What strategy/strategies did you use in order to avoid playing the 'scientist'? What worked and what didn't?

Game 7: Stick Together

1. 2 groups in total. Within each group, divide members into subgroups of 3s.

2. 1 person ties a bar magnet to a string. The other 2 members are blindfolded and will hold each end of the string and balance it.

3. They go through an obstacle course, with the first person giving directions at the start line.

Game 7: Stick Together

Key Aptitude

Emotional Quotient – Co-operation, Adaptability, Patience
Social Quotient – Teamwork, Sensitivity, Dexterity

Math/Science Concept Applied

Physics - Magnetism

Equipment/Logistics

Boxes, Paper Clips, Strings, Magnets

Time Required

30 minutes

Game Objective

To transfer the most number of paper clips with the magnet

Group Size

12-24

Procedure

1. This game is a competition between two groups. Each group will have to further divide themselves into sub-groups of threes
2. 2 of the teammates will be blindfolded while the third person known as Person A will be required to guide the 2 teammates down a designated 'course'

3. Person A will have to tie 2 strings to a bar magnet and hand each end of the string to his blindfolded teammates. (See picture)
4. The teammates will have to balance the magnet in such a manner that the magnet will not slip out of the strings
5. Person A will then guide his teammates using the commands "left, right, front, back, up and down"
6. The pair holding the magnet will then need to lower the magnet into the boxes of the paper clips along the course and transport the paper clips into the box located at the end of the course
7. After all three of them makes it to the end of the course, Person A will run back to the starting line to pass the magnet to the next trio
8. A point will be awarded for each paper clip transported
9. The group that scores the most points wins the game

Possible Variations

For extra difficulty, there could be multiple magnetic items in the box, but participants should only pick up the paper clips. Any other item picked up will not score any points

Practical Application

- How did you react to the commands given to you while you were blindfolded?
- As the one giving the commands, what were some tactics you used to make sure your intentions were understood?

Game 8: Chemical Reaction!

1. Facilitator will create an obstacle course with marbles hidden in various areas.

2. Players will pair up and try to find 8 marbles in total. They have to hold hands with their partner throughout the course.

3. Before they retrieve the marbles, they have to do a task as assigned by the facilitator.
e.g. spin around a pole 10 times

 "I'm getting dizzy!"

Game 8: Chemical Reaction!

Key Aptitude

Emotional Quotient – Communication, Motivation
Social Quotient – Teamwork, Co-operation

Math/Science Concept Applied

Chemistry – Valence Shell Theory

Equipment/Logistics

Marbles, White Tape, Tables, Chairs, Sticks, Paper

Time Required

10-20 minutes

Game Objective

To collect eight marbles through the obstacle course in the shortest time possible

Group Size

6-12

Procedure

1. An obstacle course with 8 stations will be created in an enclosed space where marbles are hidden in various areas
2. Players will play in pairs and once the games starts, all the players are to enter the room and find the eight marbles hidden amongst objects in the room

3. The pairs will hold hands and they cannot let go of each other's hand throughout the course of the game
4. Before they can retrieve the marble they spotted, the pair will have to perform a task before they can collect it. The facilitator will check on the tasks performed
5. For example, marbles can be placed beside a standing stick and an instruction sheet on the stick will inform the pair of participants to run around the stick together 20 rounds before they can collect the marble
6. The pair that finds all 8 marbles first has to make their way out of the room before they are considered winners

Possible Variations

The obstacles can be changed, such as replacing the "spinning stick" with a hop-scotch layout and so on

Practical Application

- How did you enter the course?
- What are/were good practices of a great team?
- Have we tried to move forward at the expense of leaving others behind? How did you feel if you have done so?

Game 9: Topsy-Turvy

1. Everyone lines up in 2 groups and given a plastic cup.

2. Take turns to run to the pail. Fill the cup up with water from the pail, turn it upside down, cover it with the transparency (with holes).

a) b)

3. Run to the finish line and transfer the water in another pail! Next person goes.

Game 9: Topsy-Turvy

Key Aptitude

Emotional Quotient – Patience, Persistence
Social Quotient – Dexterity, Teamwork, Co-operation

Math/Science Concept Applied

Physics – Pressure

Equipment/Logistics

Pails, Plastic Cups (with holes), Transparency Sheets, Water

Time Required

15 minutes

Game Objective

To transfer as much water as possible from one point to another

Group Size

5-10

Procedure

1. Player A from each team will run to the pail of water placed a distance away from the Start Line. With the given plastic cup, he will fill it with water from the pail and cover it with a piece of transparency
2. He will run to the finishing line with the cup upside down and transfer the water into a pail that is placed at the finishing line

3. He will then return to the starting line and the next person will continue the game
4. The team with the most amount of water transferred to the pail at the finishing line wins the game

Possible Variations

Water could be replaced with other liquids, such as oobleck (a mixture containing 1 part water & 2 parts corn-starch mixture)

Practical Application

- How did the holes hamper you from bringing water back? How did you overcome the problem/s encountered?
- What did your teammates do to help you?
- Would you have done it differently if you were to play the game again? If your answer is yes, how differently would you do it?

Game 10: Lucky 7s

1. Try to skip as many times as possible. With each skip, shout out the number of times skipped.

"3! 4!..."

2. When the number of times corresponds to a multiple of 7 or a number with the digit 7 in it, replace the count with "Lucky!".

"26, Lucky, 28..."

Game 10: Lucky 7s

Key Aptitude

Emotional Quotient - Determination, Tolerance
Social Quotient - Communication, Co-operation, Flexibility

Math/Science Concept Applied

Mathematics - Multiplication

Equipment/Logistics

Long Rope

Time Required

15 minutes

Game Objective

To skip the most number of times without anyone tripping or making mistakes

Group Size

3-15

Procedure

1. Players to stand in one straight line with two members one at each end. They are the leaders of this game
2. The two leaders are in charge of swinging the rope and all the other members are to join in the game by hopping in

3. With each hop, the players will shout out the number of times that they have skipped
4. When the group reaches a multiple of seven or a number with a seven in it, they are to shout 'Lucky' instead of the number
5. Restart the counters once someone did not skip in time or when someone shouts out the wrong number

Possible Variations

The "taboo" number can be changed

To increase the difficulty, one can also change the phrase used to a word such as "deoxyribonucleic acid"

Practical Application

- How do we make sure that no one trips?
- Was it easy to coordinate? What made it difficult for the group to win?
- How important was communication and teamwork in this game?
- Could you have succeeded without teamwork?
- How important is communication in the various aspects in your life?

Game 11: Bombs-Away!!

1. Set up the game area as shown below. Players are divided into 2 groups and each of them have a garbage bag.

pail with water pack of balloons hula hoop empty pail

2. First player runs to the pail, makes a water balloon and tosses it to the next person. The next person will catch it with the garbage bag and toss it down.

3. The last player will then burst the balloon, emptying the water in the pail.

Game 11: Bombs-Away!!

Key Aptitude

Emotional Quotient – Dexterity, Communication
Social Quotient – Co-operation, Teamwork

Math/Science Concept Applied

Physics – Estimation; Projectile Motion

Equipment/Logistics

Balloons, Pails, Water, Trash Bags, Hula-hoops

Time Required

20 minutes

Game Objective

To transfer as much water as possible from one pail to another within 5 minutes

Group Size

10

Procedure

1. Divide the group into two teams
2. Preparation: Each team will put the balloons provided and the pail of water at starting line at one end of the court

3. They will place the 4 'hulahoops' provided at an interval of 2m apart in a straight line. An empty pail will be placed beside the last 'hulahoop'. Each player has to stand in a hulahoop
4. The first player at the start line will proceed to the pail to make a water balloon. He will return to his hulahoop and toss the balloon to the next player behind him. The second player will receive the balloon with his trash bag. The water balloon will continue down the line in a similar manner
5. The last player will burst the balloon in the pail. Only one balloon is allowed to be transferred at a time

Possible Variations

Mathematics: To apply mathematics concepts, each person would be assigned a number. Each balloon will have a number written on it. The balloon can only be passed when the participant has shouted the sum/ multiplication of the 2 numbers

Practical Application

- How did you increase the efficiency of the group?
- What were your feelings when one drops the water balloon?

Game 12: Under Pressure

1. Prepare oobleck by mixing cornflour and water in the ratio of 2:1.

2. Facilitator hides some marbles in the oobleck mix beforehand.

3. Pails containing the oobleck are placed at the end of the line. Divide players into 2 groups. Participants from each group will take turns to search for marbles.

Game 12: Under Pressure

Key Aptitude

Emotional Quotient – Communication, Innovation
Social Quotient –Teamwork, Thoroughness

Math/Science Concept Applied

Chemistry – Non-Newtonian fluid

Equipment/Logistics

Cornstarch, Pails, Water, Marbles

Time Required

10-15 minutes

Game Objective

To find marbles hidden in the oobleck

Group Size

More than 4

Procedure

1. Prepare a pail of oobleck by mixing one part water to two parts cornstarch
2. Divide the players into equal groups
3. For each group, place one pail of oobleck a distance from an empty pail
4. Divide the players into equal groups

5. At the start of the race, the first player from each team will have run to the pail of oobleck and find one marble in the oobleck
6. When he has found a marble, he will run back to the rest of the team with the marble
7. The team which finds all the hidden marbles first wins

Possible Variations

Instead of finding hidden objects, the players can be asked to transfer oobleck from one pail to another

Practical Application

- How many different ways did you try out before you found the easiest method to move in the fluid?
- Did you copy another's method? If you did not, do you think your method/s worked better? Why?

Game 13: Dichotomous Frenzy

1. Divide players into groups. Ask them to each take out any item that they have.

A -wallet.   A -shoe   A -pen

2. Each group tries to make a dichotomous key based on these objects.

Game 13: Dichotomous Frenzy

Emotional Quotient – Determination, Motivation
Intelligence Quotient – Creativity, Problem-solving
Social Quotient – Self-awareness

Math/Science Concept Applied
Biology – Dichotomous Key

Equipment/Logistics
Any objects

Time Required
20 minutes

Game Objective
To categorize a group of objects by creating a dichotomous key (a step-by-step method for categorizing species)

Group Size
3-5

Procedure
1. Ask each player to place one object at the front of the play area. This object can be a personal item or anything he can find around him from the room or the surroundings
2. A personal item could be a shoe, a wallet, a pen, etc

3. The group that comes up with a complete dichotomous key for all the items at the shortest time wins the game

Practical Application

- What did you have to take into consideration so that the task could be completed quickly?
- How did the object you have used, hamper you from constructing the creation of the key? How did you overcome the problem/s encountered?

Game 14: Lost In Equation

1. Players form a line. Each player is given a card with a mathematical operation and a number.

2. Facilitator gives the 1st player a number. He will perform the operation on the card to the number and then pass the number down. Group has to try to get the correct answer!

a) Start with 4.

b) 7.

c)

Game 14: Lost In Equation

Key Aptitude

Emotional Quotient – Dexterity, Self-confidence
Social Quotient – Communication, Patience

Math/Science Concept Applied

Mathematics – Arithmetic Equations

Equipment/Logistics

Cards, Markers

Time Required

5-10 minutes

Game Objective

To get the most number of correct answers

Group Size

5-10

Procedure

1. The players will first form a line
2. The cards will be distributed to each player
3. At the start of the game, the facilitator will give the first player a number
4. The player will then use the number to perform the arithmetic operation stated on the cards and pass down the resulting number to the next player

5. This will be done until the last player finishes his arithmetic equation.
6. The facilitator will then check against the last person's answer with an answer sheet
7. The team with the most correct answers wins the game

Practical Application

- What did you feel when you were handed a tough arithmetic operation?
- How did you feel when someone took quite a while to get his answer?

Game 15: Magic 21

Facilitator will give 4 random numbers. Each student can rearrange the numbers and add any mathematical operation to yield the answer "21".

2,8,3,1!

(8+3)×2−1!

Game 15: Magic 21

Key Aptitude

Intelligence Quotient – Memory Recall, Quick Thinking

Math/Science Concept Applied

Mathematics – Mathematical Operations

Equipment/Logistics

Nothing

Time Required

5-10 minutes

Game Objective

To achieve a mathematical formula to give the number 21

Group Size

Any

Procedure

1. At the start of each round of the game, the game master will give any four single digits
2. These digits will be from 0 to 9, and any number can be repeated once in the four provided
3. Players will have to rearrange the numbers and insert their own mathematical functions to achieve the number '21'

4. The fastest person to formulate an accurate equation will shout out his/her answer

Possible Variations

More than four digits can be provided for each game round

Numbers may be repeated more than once for each round

Instead of a total number of 21, the game master may choose any other number as long as the initial numbers provided are sufficient to meet the final number

Practical Application

- What kind of challenge did you face when you have difficulties coming with the equation?
- How did you overcome the difficulties?

Game 16: Tumble Dry

1. Have one player stand in the center of a circle formed by the rest of the players (their shoulders should be touching each other).

center guy

2. The player in the center stands firm with his feet together and arms crossed over his chest.

3. Within the circle, the center player will lean towards the side and be "pushed" left or right.

Game 16: Tumble Dry

Key Aptitude

Emotional Quotient – Self-confidence
Social Quotient – Trust, Teamwork, Co-ordination

Math/Science Concept Applied

Physics – Forces

Equipment/Logistics

Nothing

Time Required

5-10 minutes

Game Objective

To move a selected player back and forth according to the instructions of the game master

Group Size

6-8

Procedure

1. Shoulders touching shoulders, the whole team will form a circle
2. One player will stand in the centre of the circle whilst the rest of the members close the gap to form a smaller ring
3. The player in the centre needs to stand with his feet together with his arms crossed over his chest

4. When the game master starts the game, the player in the centre will lean towards one teammate without bending the body
5. The players will listen to the game master's instructions and push the centre player back and forth or left and right
6. With the support of the whole team, the player in the centre should not fall at any point in the game

Possible Variations

There can be two players in the centre of the circle being pushed back and forth

Practical Application

- As the centre player, how did you feel?
- Did you trust your team-mates to keep you safe without falling?
- What do you think will happen if your team-mates do not pay attention and you get injured?

Game 17: Blow Wind Blow

1. Everyone sits in a circle on chairs, except for one person who will stand in the middle of the circle. They will each adopt the identity of a biological organism/object.

2. Person in the middle will announce a property. Anyone whose identity fits the property will have to sit at another chair. The person in the middle also tries to find a place to sit.

Developing Life Skills Through Math and Science Games

Game 17: Blow Wind Blow

Key Aptitude

Emotional Quotient – Alertness, Attentiveness, Honesty
Social Quotient – Spontaneity

Math/Science Concept Applied

Biology – Organisms

Equipment/Logistics

Nothing

Time Required

20 minutes

Game Objective

To find a seat as soon as possible

Group Size

20-30

Procedure

1. Every player in the group will sit in a circle on chairs provided except for one player
2. The facilitator will give out pieces of paper with different biological terms or organisms to each player (including the one in the centre)
3. Without telling each other what they are, the player in the centre will need to find a seat by 'blowing someone away'

4. The player in the center can make statements pertaining to characteristics of different organisms or biological terms
5. For example, he can say 'Blow all of you who have 2 legs', and anyone whose stated organism has 2 legs must find another seat
6. The game continues on and the player standing in the centre at the end of 5 minutes loses

Possible Variations

Chemistry: To apply Chemistry concepts, the biological terms could be replaced with chemical common compounds, and statements that are shouted will pertain to the properties of the chemical (such as if they are soluble in water)

Practical Application

- Were you tempted to cheat since nobody knew the organism you represent?
- How did you feel when you were ousted out to the center of the circle of chairs?

Game 18: Splish, Splash, Splosh

1. Divide players into 2 groups. Within each group, number participants from 1,2,3, etc. They stand in a row, holding a pail and scoop each, back-faced to the other person with the same number.

2. Facilitator calls out a math equation. The person whose number corresponds to the answer has to turn around and splash water onto his counterpart before he gets splashed.

$5^2 - 21$!

"gotcha!"

Game 18: Splish, Splash, Splosh

Key Aptitude

Emotional Quotient – Agility, Responsiveness
Social Quotient – Teamwork, Co-operation

Math/Science Concept Applied

Mathematics – Arithmetic Formulas

Equipment/Logistics

Pails, Scoops, Water

Time Required

20 minutes

Game Objective

To react as fast as possible so as to earn points for the team

Group Size

10-20

Procedure

1. Choose a gamekeeper if there are odd number of people participating (choose 2 if there are even number of people playing)
2. Divide the participants into 2 teams
3. Each team will line up in a straight row parallel to the other team
4. Each pair facing each other will be given a number starting from 1 and increases as the group goes down the line

5. Every player will have a pail full of water and a scoop
6. Every player will turn around such that his back faces the opponent's
7. The game keeper will shout a mathematic statement involving symbols like +−×÷ (e.g. 2+3)
8. Each participant will have to calculate the answer
9. If the answer tallies with the number allocated to the participant, the participant turns around and scoop a bit of water and splash it on the opponent who would be doing the same
10. As he splashes the water, he has to shout "gotcha"
11. The gamekeeper has to decide who is faster
12. If Participant A is faster, his team will gain 10 points
13. If the participant splashes the water when it is not supposed to be his turn, 5 points will be deducted from the team
14. The gamekeeper may choose to increase the difficulty of the questions (e.g. what are the prime factors of 12?) and there could be 2 answers to the question thus two pairs of participants will splash at each other instantaneously
15. After 15 minutes, the team with the highest score wins the game and the losing team has to do a forfeit

Practical Application

- As a participant, how were you challenged when the numbers were changed?
- How did you adapt to the changes?

Game 19: Reflective Lights

1. In a dark room, everyone stands in a zigzag manner, holding a mirror. The person at the end holds a torch.

2. When the torch is switched on, players will try to tilt the angle of their mirror, letting the light reflect off till it reaches the last person.

"how should I tilt this..."

Game 19: Reflective Lights

Key Aptitude

Emotional Quotient – Agility, Co-ordination
Social Quotient – Communication, Teamwork

Math/Science Concept Applied

Physics – Reflection of light

Equipment/Logistics

Mirrors, Torchlight, Stopwatch, Dark Room

Time Required

10 minutes

Game Objective

To direct the beam of light to a specific destination in the shortest time possible

Group Size

10

Procedure

1. All players will align themselves in a zigzag manner, each holding a mirror. The first player will hold a torch
2. Once the torchlight is switched on, the players will change the angle of their mirrors such that the light will be directed to the last player holding the mirror

3. The timing is stopped once the last player received the beam of light and bounced it off his mirror
4. The group may attempt more complicated formations or standing arrangement like forming a heart shape such that the last person reflect the light to the first player holding the mirror

Possible Variations

Laser Pointers could be used, but participants should be given protective eyewear gear (such as goggles) as a precaution

Practical Application

- What could have been done to improve the speed of getting things done?
- What scientific concepts did you learn from playing this game?

Game 20: Get Into Shape

Facilitator will name a shape. The group has to try to form as many of that shape as possible using their bodies as props.

Game 20: Get Into Shape

Key Aptitude

Emotional Quotient – Dexterity
Social Quotient – Communication, Teamwork

Math/Science Concept Applied

Mathematics – Geometry; Shapes

Equipment/Logistics

Nothing

Time Required

20 minutes

Game Objective

Using only the body of each player in the team to form as many shapes as possible

Group Size

Any (preferably even number)

Procedure

1. Each team should have equal numbers of players
2. When the facilitator names a shape, the players will have to form the same shape, using only their bodies
3. The team with the most number of the named shape wins

Practical Application

- What were the challenges faced when the facilitator changed a certain aspect of the game?
- How did you help your teammates to get into the 'desired' shape?

Game 21: Who's Walking Now?

Players are each given an animal to act out. They have to walk like the animal from the start to finish line. Group members try to guess what animal it is!

Game 21: Who's Walking Now?

Key Aptitude
Emotional Quotient – Innovation, Spontaneity
Social Quotient – Communication, Teamwork

Math/Science Concept Applied
Biology – Biodiversity

Equipment/Logistics
Container, Paper, Markers

Time Required
20 minutes

Game Objective
To reach the finish line walking as an animal

Group Size
20-30

Procedure

1. The facilitator will split the players into groups
2. A game moderator will be required for each group
3. At the start line, a container of different animal names is written on each piece of paper

4. When the game starts, the first player of each group will pick a piece of paper from the container and move like the 'animal' stated on the paper towards the finish line
5. Once he/she has reached the end, the group members are to guess the animal
6. They are given 3 chances to guess the animal. If the group gets it wrong, the first player will have to run back to the start line again and pick out another piece of paper (another animal named)
7. If the team guesses the second animal correctly, the player who has reached the finish line can stay there and make guesses for the team
8. The game continues when another player repeat steps 4,5 & 6

Practical Application

- As a participant, how were you challenged when you had to assume the 'animal' you had picked?
- Describe some of the fun moments when you wrongly guessed the 'animal' that your teammate assumed?

Game 22: Long and Short

Facilitator will state a measurement. With the help of a measuring tape, players will use items around the room and piece it together to achieve that length.

~50 cm

"I wonder how long this is now."

Game 22: Long and Short

Key Aptitude

Emotional Quotient – Co-ordination, Spontaneity
Social Quotient – Communication, Teamwork

Math/Science Concept Applied

Physics – Measurements

Equipment/Logistics

Measuring tapes

Time Required

20 minutes

Game Objective

To be the first group to achieve the desired 'lengths' using different objects

Group Size

10-20

Procedure

1. The game is played in pairs or in groups of 'threes'
2. The facilitator will give each team a measuring tape
3. When the players are ready, the game master will state a certain measurement
4. Using items found around the room, the players are to find different objects that fit the measurement required

5. These objects must be measured in whole
6. For example, the facilitator can state 20 cm, and this can be achieved with the 4 legs of the chair or 2 pens laid side by side or lengthwise

Possible Variations

The facilitator can distribute kitchen weighing scales and use weight of objects instead of lengths

Practical Application

- In what way did your fellow group members help you?
- Could the communication between team members be improved? In what ways could it be improved?

Game 23: Bumper Particles

1. Players start off anywhere in the room, facing any random direction, walking in a straight line.

a) Change direction if you collide with someone else.

b) Change angles if you collide into another object.

2. The facilitator will stand at the corner of the room. Treat him like an object and try to "bump" him to the other end.

Facilitator

"bump!"

Game 23: Bumper Particles

Key Aptitude

Emotional Quotient – Goal-setting
Social Quotient – Adaptability

Math/Science Concept Applied

Chemistry – Brownian Motion

Equipment/Logistics

Nothing

Time Required

10-20

Game Objective

Move the 'Dust' from one corner of the room to another

Group Size

30-50

Procedure

1. The whole group of players will move randomly around the room in a straight line and changing direction or angles only when they are 'bumped' by a another player or if they 'bump' into any object in the room
2. The players are to move the facilitator from one corner of the room to another

3. The facilitator can only move when he is 'bumped' by the 'gas particles' (i.e. the players)

Possible Variations

A heavy object like a huge cardboard box can be used to represent the 'dust' particles

Practical Application

- How did you keep moving in order to achieve the goal of moving the facilitator?

Game 24: Bouncing Balls

1. Set up the room as follows. 2 members stand outside, other members stand anywhere inside.

☐ table
X players from group 1
O Players from group 2

2. Facilitator says a number from 1 to 4. The players then needs to bounce a table tennis ball off the tables for that number of times and have their groupmates catch it.

,2!

Game 24: Bouncing Balls

Key Aptitude

Emotional Quotient – Dexterity, Determination
Social Quotient – Co-ordination, Communication

Math/Science Concept Applied

Physics – Projectile motion

Equipment/Logistics

Tables, Ping Pong balls

Time Required

10 minutes

Game Objective

To get as many ping pong balls to your team members as possible

Group Size

8-10

Procedure

1. Place 16 tables (4 by 4) in the middle of the room, leaving about 30 cm spacing between each of the tables
2. Each team consists of 4 or 5 players
3. 2 players (Catchers) will stay outside of this table arrangement while the rest of the team can move around the inside of the arrangement. These

catchers have to be careful not to shift the tables around when walking inside the area

4. When the game starts, the facilitator calls out a number from 1 to 4

5. The number that is called out will be the number of bounces the ping pong ball needs to bounce off the table. This is executed by the throwers who are standing outside the table formation to the catchers who are inside it

6. For example, if the facilitator calls out 3, the thrower must throw the ping-pong ball onto a table, bounce it off a second and third table before the catcher from the same team can catch the ball

7. If the ball does not bounce enough times, it will not be counted.

8. Likewise, if a ball bounces twice if the facilitator says 'one' the ball will not be counted as well

9. The teams with the most number of ping pong balls at the end of the game wins

Possible Variations

To increase the difficulty of the game, tables can be moved further away from each other and 2 teams can be playing the game at any given time with 2 different colours of ping pong balls to differentiate the teams.

Practical Application

- How does teamwork help in the game you just played?
- How did you deal with the situation when it became frustrating?

Game 25: Circle of Trust

1. Everyone holds hands and stand in a circle inside a hula hoop.

←Hula hoop

2. The whole team needs to get under and out of the hula hoop without letting go of their hands.

"Try to move the hoop up with your legs!"

Game 25: Circle of Trust

Key Aptitude

Emotional Quotient – Co-operation, Trust
Social Quotient – Communication, Co-ordination

Math/Science Concept Applied

Biology – Quadrupendalism

Equipment/Logistics

Hula hoops or Strings (to form a circle)

Time Required

20-30 minutes

Game Objective

To get under the 'hole' as a team

Group Size

10-20

Procedure

1. Everyone in the team must stay in the circle inside a hulahoop or a circle made from a string and hold hands with each other
2. The objective is for the whole team to get out of the circle from underneath the hulahoop or string
3. The players cannot use their hands, arms or shoulders to lift the circle, only their lower limbs can touch the circle

Practical Application

- What could have been done to improve the speed of getting things done?
- In what situation/s in the game did you feel the need to give encouragement to others in the team?

Game 26: 360° Water

1. Set up the game area as follows:

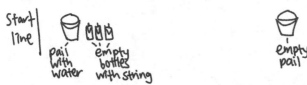

2. Players will take turns to fill the empty bottles with water, swing it while running to the empty pail and empty its contents in the pail.

3. Try to fill the pail up as quickly as possible!

Game 26: 360° Water

Key Aptitude
Emotional Quotient – Alertness, Determination
Social Quotient – Teamwork

Math/Science Concept Applied
Physics – Centripetal Force

Equipment/Logistics
String, Pails, Small Plastic Bottles, Water

Time Required
10-30 minutes

Game Objective
To fill up the pail with as much water as possible in the shortest possible time

Group Size
10-20

Procedure
1. Facilitator will tie the string to the top of the plastic bottles
2. He will place a pail filled with water at the start line and an empty pail at the finish line
3. The players will be divided into teams of about 5-8 players each

4. When the facilitator starts the game, the first players of each team are to fill the small plastic bottles with water
5. The rule for moving from the start line to the finish line is that the players must swing the bottles at 360° at all times as they walk or run to the empty pail
6. Upon reaching the pail at the finish line, he will empty the water from the bottle into the pail and run back to his/her teammates and pass the bottle to the second player of the team
7. The second player of the team can then fill up the bottle and start moving to the pail at the finish line
8. The team with the most amount of water in the pails at the end of 10 minutes wins the game

Practical Application

- As teammates, how did you help others to overcome the 'difficulties' faced?
- What were some factors that determined the success of the team?

Game 27: Knotted DNA

1. Players are paired up and given a piece of rope. Try to create knots that are as complicated as possible.

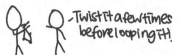

2. After 2 minutes, all ropes will be collected by the facilitator and redistributed. Players have to try and untie the knot as quickly as possible.

Take this.

Game 27: Knotted DNA

Key Aptitude

Emotional Quotient – Dexterity, Patience
Social Quotient – Co-operation, Teamwork

Math/Science Concept Applied

Biology – Genetics

Equipment/Logistics

Pieces of Rope

Time Required

10-15 minutes

Game Objective

To entangle the rope

Group Size

10-20

Procedure

1. Players are paired into teams of 2 and given a piece of rope
2. They are to create the most complicated knots and twists in the rope within 2 minutes
3. Once the 2 minutes are up, the game master will collect back all the pieces of rope and distribute them randomly to the pairs

4. Once the players are ready, the team will straighten out their piece of rope and the first team that does so, wins the game

Practical Application

- What were your feelings like when you could not straighten out the rope?

Game 28: Cosmos!

1. Form groups of 5-8. Each person has a slinky but all slinkys in the group are entangled beforehand. The group has to try to untangle them.

"Move this to the left..."

2. Use their bodies and slinkies to create the largest enclosed area possible!

Birds-eye view:

"Maybe you should move there!"

Game 28: Cosmos!

Key Aptitude
Emotional Quotient – Dexterity, Communication
Social Quotient – Co-operation, Teamwork

Math/Science Concept Applied
Physics – Big Bang Theory

Equipment/Logistics
Slinkys, Measuring Tape

Time Required
10-20 minutes

Game Objective
To separate the Slinkys and form the largest structure

Group Size
20-30

Procedure
1. Form equal groups of 5-8 players
2. There must be the same number of Slinkys as players in the group
3. The facilitator will entangle all the Slinkys for each group and place the bundle on the floor
4. The players are to form a circle around the entangled Slinkys

5. When the players are ready, the facilitator will start the game and the players have to work fast to separate each Slinky in the mess and then creatively use their own bodies and all the 'Slinkys' to connect together to create the largest enclosed area (Universe)
6. After 10 minutes of game play, the facilitator will measure the area of each galaxy and determine the largest one which wins the game

Possible Variation

Instead of using Slinkys, pieces of elastic strings or skipping ropes can be used

To enhance the difficulty of the game, the facilitator can instruct that only one player speaks throughout the game or only one hand of each player can be used

Practical Application

- What could have been done to improve the speed of getting things done?
- How does teamwork help in the activity you just took part in?
- Were too many people shouting orders? If so, what should have been the correct Procedure?

Game 29: Seeds and Sticks

1. Players line up, leaving some space in between. The 1st player is given a term and has to use only saga seeds to convey the term to the next player without spelling it out.

2. The next player will then use only sticks to convey the term to the 3rd person. The 3rd person will use saga seeds, and so on to pass the message down the line.

3. The last person tries to guess what the word is based on the diagram of the 2nd last player.

Game 29: Seeds and Sticks

Key Aptitude

Emotional Quotient – Social Awareness, Empathy
Social Quotient – Teamwork, Communication

Math/Science Concepts Applied

Biology – Any biology terms or concepts taught (e.g. Cell Biology)

Equipment/Logistics

Saga Seeds, Sticks, Table

Time Required

30 minutes

Game Objective

To complete the game as quick and accurate as possible

Group Size

Any even number (but preferably less than 20)

Procedure

1. A moderator is appointed to keep track of the scores of the two teams and to ensure that the teams follow the rules of the game
2. Sticks and saga seeds will be placed into several different piles on each table
3. The moderator will stand in the middle of the two tables facing the audience while all the players sit on the floor together

4. Within each team the leader will decide on the order in which the players will go.
5. The first player from both teams will get a word from the moderator like 'enzymes' or 'vascular bundle'
6. The 2 players will each stand at a table and the next player will proceed to the tables
7. The first player will have to use only the given saga seeds to illustrate the word in a diagram form. He is not allowed to spell out the word/words. The second player could only guess the word without saying it/them out. Before the first player leaves the table, he must gather all the seeds into a pile again
8. The second player will use only the sticks provided to convey the message to the third player. When it is the third player's turn, the second player must gather all the sticks into a pile on the table
9. The third player will only use saga seeds to convey his message
10. The game continues until the last player's turn. He will tell the moderator what the 'word' is
11. The moderator will decide if he gets the correct answer and awards the score. If the answer is wrong, the last player will continue guessing until he gets the correct answer
12. If the last player gives up, no scores will be awarded

Possible Variations

Physics/Chemistry: Topics used in the game can be from any subject topic

Practical Application

- How did you feel when the player could not guess the word you were trying to convey?
- How did you overcome your feelings of frustration?

Game 30: Water Waves

1. Each pair of players will hold a trash bag. Stand in a circle and pass the water balloon to the next person while arranging themselves according to height.

water balloon

trash bag

"You are taller than them, swap places!"

2. When that is done, break the circle and pass the balloon to the tallest pair.

Game 30: Water Waves

Key Aptitude

Emotional Quotient – Communication, Dexterity
Social Quotient – Teamwork, Co-operation

Math/Science Concept Applied

Physics – Hydrodynamics

Equipment/Logistics

Balloons/Plastic Bags, Water, Trash Bags

Time Required

15-25 minutes

Game Objective

Be the first group to line up according to height while concurrently passing balloons that are filled with water

Group Size

5-10

Procedure

1. Group participants will first form pairs, with each pair holding a trash bag
2. The pairs will then form a circle and a balloon filled with water will be passed around in a clockwise manner
3. Each pair should take no longer than 2 seconds to pass the balloon to the next pair

4. The participants will get to practise 'the passing movement' in a circle, after which they will be instructed to line up according to height from the shortest to the tallest while passing the water balloon

Possible Variations

The number of balloons used can be increased from one to two or three.

The sequence of arrangement could be changed from height to other details, such as foot size or arm length (anything physically visible)

Practical Application

- How difficult was it to change positions with one another?
- How did your group manage to accomplish the task while trying to pass the balloon around?
- When and how did the team decide that the group has accomplished the task?
- What was learnt about communication during this game?
- When working as a team, is it right to keep thinking that your opinions and solutions are always right?
- What importance does a leader play in similar situations?

Game 31: Living Dominoes

1. Use as many dominoes as possible to make different types of cells.

dominoes

2. Label the cell parts and try to form specialised cells!

Red blood cell,
biconcave shape

Game 31: Living Dominoes

Key Aptitude

Emotional Quotient – Patience, Motivation
Intelligence Quotient – Memory Recall, Application
Social Quotient – Teamwork, Communication

Math/Science Concept Applied

Biology – Cells

Equipment/Logistics

Dominos, Markers, Paper

Time Required

45 minutes

Game Objective

To earn as many points as possible for the group

Group Size

10-25

Procedure

1. Players will use the dominos provided to form the structure/s of any type of cells
2. After arranging the structures of the cells, they have to label the various parts of the cells with pieces of paper provide

3. More points will be awarded for forming 'specialized' cells with distinctive features and be able to state the various adaptations of the cell (e.g. red blood cells)
4. Players should use up as many dominos as possible but they have to make sure that the dominos will topple
5. 1 point will be awarded for each domino toppled and for correct labelling. If all the dominos topple, an additional of 20 points is added. If the group chooses to arrange a specialized cell, an additional of 20 points is awarded. Correct description and explanation of the specialized part/s will give the group 2 points

Possible Variations

For a younger age group, one could also form shapes of plants, animals or anything else made up of cells

Practical Application

- What were the difficulties you experienced while forming the structure/s?
- Were there any agreements or disagreements in your group? How did you solve the disagreements?
- How would you solve them better the next time round?

Game 32: Chemical Creation

1. Label balloons with different element names.

2. Facilitator announces a compound. Try to form it using the balloons and string.

1 string for single bonds   2 strings for double bonds

Game 32: Chemical Creation

Key Aptitude

Emotional Quotient – Adaptability, Persistence
Intelligence Quotient – Memory Recall, Application
Social Quotient – Teamwork, Communication

Math/Science Concept Applied

Chemistry – Organic compounds

Equipment/Logistics

Inflated Balloons, Strings, Markers

Time Required

45 minutes

Game Objective

To assemble a structural formula as fast and accurately as possible

Group Size

5-15

Procedure

1. In two different groups, the participants will use the markers provided to write the symbols of the elements which are parts of the compound on balloons that are provided. (e.g. write carbon as "C")
2. Participants in the group will arrange the balloons to form the structural formula of the organic compound as called out by the facilitator

3. They then tie the balloons together with a string to show that it is a single bond, or tie 2 strings between the balloons to represent a double bond
4. 30 points will be awarded for a correct arrangement with no mistakes. With each mistake, deduct 5 points from the 30 points and then award that number of points to the group
5. An additional 10 points will be awarded if the group can state the homologous series of the compound
6. The group with the most number of points after 45 minutes wins the game

Possible Variations

Use the least amount of time to form an organic compound

Biology: To apply biology concepts, one could also change the objective to forming base pairing in a stipulated DNA sequence (AT and GC pairs)

Practical Application

- How did you link up the balloons while making sure they did not float away?
- What roles did everyone in the group have to play?

Game 33: Blast-Off!

1. Each group is given a 500-ml bottle. They can put any composition of chemicals inside the bottle.

2. They are then given baking soda and vinegar to put in the bottle. Try to make its contents fly as far as possible.

Developing Life Skills Through Math and Science Games

Game 33: Blast-Off!

Key Aptitude

Emotional Quotient – Innovation, Communication
Social Quotient – Teamwork, Co-operation
Intelligence Quotient – Problem-solving, Application

Math/Science Concept Applied

Chemistry – Acid and Bases

Equipment/Logistics

Tables, 500ml bottles, Vinegar, Baking Soda, Paper

Time Required

30 minutes

Game Objective

Design the bottle such that when vinegar and soda powder are added, the bottle would "fly" the furthest

Group Size

Minimum 2

Procedure

1. Divide the group into two groups and distribute a 500ml bottle to each group

2. Each group has to decide what to add to the bottle such that it could travel the furthest, taking into consideration elements like weight and air resistance
3. After 20 minutes, each group will decide the amount of vinegar and baking soda to be used and then placed them in the bottle, shake it and let the item fly!

Possible Variations

Baking soda and vinegar can be replaced by coke and sugar-based candy

Practical Application

- What could you add to the bottle to overcome air resistance?
- What did you have to take into consideration so that the distance travelled could be maximized?

Game 34: Losing My Marbles

1. Players each have a board and line up in order of height. Have a bowl of marbles beside the 1st player and an empty bowl beside the last player.

2. Pass the marbles from one bowl to the other, only using the boards, without letting them drop!

Game 34: Losing My Marbles

Key Aptitude

Emotional Quotient – Communication, Dexterity
Intelligence Quotient – Problem-solving
Social Quotient – Teamwork, Co-operation

Math/Science Concept Applied

Physics – Balance

Equipment/Logistics

Flat boards (at least 30 cm in length), Marbles, Bowls

Time Required

10 minutes

Game Objective

To pass the marbles to every player in the room

Group Size

More than 10 players (2 groups)

Procedure

1. Facilitator will divide the players into 2 groups
2. He will ask all participants to line up in terms of height, from the tallest to the shortest
3. 20 marbles will be put into a small bowl and placed beside the first player

4. The players must pass the marbles from one to another in the shortest time possible using only the flat-boards provided without touching the marbles with their hands
5. Once the game starts, the first player will place a marble onto his board and pass the marbles to the next player without dropping any marbles
6. The last player will then have to direct the received marble into a bowl placed on a chair beside him
7. The team with the most marbles in the bowl at the end of the line wins

Possible Variations

More marbles or balls can be placed on the flat surface

Practical Application

- What kind of difficulty or frustration did you encounter whilst trying to balance the marbles?
- How did you help your teammates to get the task done?

Game 35: Virus Attack

1. The game is played on a tiled floor. Players need to try and go from one end of the tiles to another without stepping on "invisible mines" on the tiles (determined by facilitator beforehand).

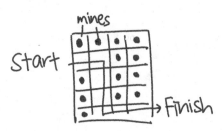

2. The 1st time a player steps on a mine, he will be informed, but the 2nd time he steps on one, he will be "immobilized".

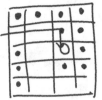

"You are immobilized!"

Game 35: Virus Attack

Key Aptitude

Emotional Quotient – Memory Recall, Determination
Social Quotient – Co-operation, Teamwork

Math/Science Concept Applied

Biology – Immunity

Equipment/Logistics

Tiled Floor

Time Required

15-30 minutes

Game Objective

To move all the players from the group to the other side of the play area in the shortest time

Group Size

6-10

Procedure

1. Facilitator will map out 'invisible mines releasing an immobilizing virus' on certain square tiles in the room
2. Only the facilitator will know this map and the locations of the invisible mines
3. One player from each group will be allowed to move at any one time

4. The player will have to move from one tile to another without stepping on the 'virus filled mine'
5. If one player steps on a 'mine', the facilitator will shout 'infected', and the player can step back onto a safe tile
6. The team will have to mentally remember the locations of these mines
7. If the player steps on a second mine, the facilitator will shout 'immobilised'
8. The player is immobilized by the virus and would need to be 'rescued' by a player who has completed the course. This player in the team will use 'the antidote' located at the end of the room to save the player who is immobilized by tapping on his shoulder
9. Once a player stops moving, another team member can start to move from the start line, taking care not to step on the safe tiles only
10. The game ends when all the team members have reached the finish line

Possible Variations

To increase the difficulty, the game master could have 2 maps showing the location of the mines, and alternate between the two every time a participant moves.

Practical Application

- How did you feel when you were immobilized?
- What were your feelings like when you were rescued?
- How did rescuing your teammate/s help you reached your team's goal/s?

Game 36: Let's Jam!

1. Divide players into groups of 3. Each of them will be a receiver, transmitter and jammer. The receiver is blindfolded. Lay out poker cards on the floor.

Transmitter, Jammer

Receiver

cards

2. The transmitter will try to ask the receiver to pick up the cards in a certain order, while the jammer will try to disrupt the process.

"Pick up the left card!"

"No, the right!"

Game 36: Let's Jam!

Key Aptitude

Emotional Quotient – Sensitivity, Interpretation, Persistence
Social Quotient – Communication, Teamwork, Sensitivity

Math/Science Concept Applied

Physics – Analogue Transmissions

Equipment/Logistics

Poker Cards (One complete suite), Blindfolds

Time Required

5-10 minutes

Game Objective

To pick up the poker cards according to a stipulated order in the fastest time possible

Group Size

6-12

Procedure

1. Facilitator first spreads the suite of cards evenly on the floor
2. 3 players will be chosen to fulfill the following roles: A receiver, a transmitter and a jammer

3. At the start of the game, the facilitator will tell the transmitter how the receiver should pick up the cards (such as from the smallest value to the biggest)
4. The receiver who is blindfolded will be led to the cards
5. Once the game commences, the transmitter will give directional commands to the receiver to pick up the cards
6. Concurrently, the jammer will try to confuse the receiver by giving different directional commands
7. The winner will be the team who can pick up the cards in the fastest time possible

Possible Variations

There could be an additional receiver picking up the cards

Another suite of cards could also be added

Practical Application

- As the transmitter, how did you get your message across to the receiver? What method/s worked and what did not?
- As the receiver, how did you distinguish between the different commands given?

Game 37: One Blind Mouse

1. The team stands at the end of the room. One of them is blindfolded.

obstacles

2. Equation is written on the board. Group members have to try to direct the blindfolded member to the board and plot the graph.

"you are reaching the board soon!"

Game 37: One Blind Mouse

Key Aptitude

Emotional Quotient – Honesty, Persistence
Social Quotient – Communication, Teamwork

Math/Science Concept Applied

Mathematics – Graphs

Equipment/Logistics

Large Whiteboard or Chalkboard

Time Required

10-15 minutes

Game Objective

To draw a graph blindfolded

Group Size

More than 3

Procedure

1. The whole team will be asked to stand at the end of the room, facing the whiteboard
2. The team will be asked to select one player to draw a chart or graph on the whiteboard while the rest of the team are to remain in their position

3. Once the selected player's eyes are blindfolded, the game master will write a simple mathematical formula on the whiteboard, large enough for the whole team to see them
4. With the help of the team, the selected player will have to walk to the front of the class and draw the graph according to the given formula
5. Once the graph is completed, the player will have to go back to the end of the room and the rest of the team will continue with the game

Possible Variations

Some furniture can be placed in the room to act as obstacles once the player has closed his eyes

To increase the level of difficulty, the game master may want the players to mark a point in the diagram by writing a value on the whiteboard

Practical Application

- Was it difficult to follow the instructions shouted out by so many players in your team at the same time? What were some of your frustrations? How did you overcome them?
- Was it difficult to be honest while your eyes were blindfolded?

Game 38: That Sinking Feeling

1. Facilitator prepares a pail of saltwater, freshwater and a glass cylinder with a floating egg in it as a model.

saltwater   freshwater   mix of freshwater and saltwater

2. Players are divided into 2 groups. Each group is given 3 empty glasses, 3 quail eggs and a paper cup with holes in the bottom.

$$1 \text{ group} = 3 \times \underline{\bigsqcup} + 3 \times \text{☺} + \text{◻}$$

3. Players use the paper cup to collect the freshwater, saltwater or a mix. They then run to the finish line, pour it in the glass, drop the quail egg in and try to align it with the model.

Game 38: That Sinking Feeling

Key Aptitude
Emotional Quotient – Intuition, Perceptibility
Social Quotient – Teamwork, Co-ordination

Math/Science Concept Applied
Chemistry – Density

Equipment/Logistics
Tall and Thin Glass Cylinders, Quail Eggs, Salt, Water, Paper Cups (with Holes)

Time Required
15 minutes

Game Objective
To align all 3 eggs within 3 cylinders as shown in the model in the fastest time possible

Group Size
5-10

Procedure

1. Prior to the start of the game, the facilitator will prepare a pail of salt water and another pail of fresh water. These pails are to be labeled clearly
2. The facilitator will also prepare one glass cylinder with a floating quail egg inside as the 'model'

3.   The players are divided into two groups
4.   Each group is given a paper cup with holes at the bottom
5.   Before the start of the game, the players are told to attain the same floating height of the quail egg for their 3 empty glasses similar to that of the facilitator's model
6.   To do this, they are given water, salt water, 3 quail eggs and a paper cup with a hole at the bottom. These are placed at the start line.
7.   The players will have to bring water to the finish line using the paper cup provided.
8.   They can take either fresh water or salt water
9.   The players can use the 'trial and error' method, throwing away the water in the glass cylinders if they are not able to attain the desired result which is a mix of fresh and salt water
10. Once they are satisfied, they can drop in the quail egg
11. The team that gets all 3 cylinders of eggs aligned to that shown in the model cylinder wins the game

Possible Variations

Instead of salt, other chemicals that can be dissolved in water could be used

An alternative to increase the difficulty would be using simple syrup (which is made of 1 part water and 1 part sugar)

Practical Application

- How did you adapt to the sudden changes during the game?
- In what way/s did your fellow group members help you?

Game 39: Categorize Me!

1. Players are divided into groups. Each group is given one paper plate and will have some random items placed at the end of the room.

randomitems

2. Facilitator will call out a property. One player from each group will go to the table, find an item that fits that property, place it on the plate and balance it back.

3. Repeat with different phrases and properties!

Game 39: Categorize Me!

Key Aptitude
Emotional Quotient – Responsiveness, Dexterity
Social Quotient – Teamwork, Co-ordination

Math/Science Concept Applied
Biology – Classification

Equipment/Logistics
Any Items (e.g. Batteries, Wires, Balls, Plates, Forks, Paper), Paper Plates

Time Required
20 minutes

Game Objective
To differentiate properties of various materials

Group Size
10-20

Procedure
1. The facilitator divides the players into 5 even groups
2. He gives a paper/plastic plate to each group
3. All the other items are then placed into one large container at one end of the room
4. The groups of players will line up at the other end of the room.

5. When the players are ready, the game master will shout the property of desired items (e.g. I can conduct electricity)
6. The first person of each group will have to run to the container and find an item which conducts electricity. He will place the item on the plate and balance the plate on his head, making his way back to his team
7. If the plate drops, the player will need to return to the large container and start balancing the plate from there again. The player who reaches his group first with the correct item gets a point
8. Once the players are back in place, the game master can shout another phrase
9. The second player will need to run back to the container to return the first item and find the next item with the desired property
10. While the player must balance the item on his head (with the help of the paper plate) when moving from the container to his group, he holds the item and plate in his hand when leaving his group to go to the container of items
11. The group with the most points at the end of the game wins

Practical Application

- What kind of difficulty or frustration did you encounter whilst trying to balance yourself?
- As teammates, how did you help others to overcome the 'difficulties' faced?

Game 40: Rhythm of Life

Do a skit based on the biology topic assigned without saying the word itself! Audience will try to guess the topic!

Game 40: Rhythm of Life

Key Aptitude
Emotional Quotient – Innovation
Social Quotient – Teamwork, Problem-solving

Math/Science Concept Applied
Biology – Various topics

Equipment/Logistics
Pails, Envelopes, Paper, Markers

Time Required
1 hour

Game Objective
To come up with a performance revolving around the biology topic as vividly as possible

Group Size
10-30

Procedure
1. Each group of 5 will choose an envelope that contains the biology topics
2. They will have to come up with a 5-minutes performance (either in the form of skit/dance/song) revolving around the topic chosen within 15 minutes, without voicing out the biology topic

3. The group will have to perform in front of everyone and everyone gets to guess the topic the group had been handed
4. The group of audience (the other group members), which guessed the correct topic with the least number of guesses, wins the game

Possible Variations

Chemistry/ Physics/ Mathematics: Chemistry, Physics or Mathematical topics may be used instead of Biology

Practical Application

- What did you feel when your group was handed a tough topic?
- How did you feel when the group took quite a while to get the performance worked out? How did you manage your emotions?

Game 41: Mini-TV!

1. Each person is given a graph paper, with specified dimensions.

4x4

2. Facilitator has an image in mind. He will say a series of "yes" and "no", depending on whether it is "colored" or not.

_No, yes, yes, No ...

3. Players can shade and try to guess what is the object the facilitator is describing.

_A circle!

Game 41: Mini-TV!

Key Aptitude

Emotional Quotient – Communication, Persistence
Social Quotient – Adaptability

Math/Science Concept Applied

Physics – Electron Beam Scanning

Equipment/Logistics

Pictures drawn on Graph Paper, Stationery

Time Required

15-20 Minutes

Game Objective

The one who guessed the most number of pictures correctly wins

Group Size

5-10

Procedure

1. Participants will be given blank pieces of graph paper
2. The facilitator will then "define" the drawing area by naming the dimensions of the drawing at hand (e.g. 5 by 5)
3. Once the participants are ready, the facilitator first gives a hint on what is being "scanned"

4. The facilitator will "scan" the drawing, starting from the top left of the graph paper to the bottom right, saying either 'yes or no'. 'Yes' will refer to a square to be shaded, while 'no' refers to a 'blank' square

e.g. For a circle; Hint: A Shape

The facilitator would say No, Yes, Yes, Yes, No, Yes, No, No, No, Yes, Yes, No, No, No, Yes, Yes, No, No, No, Yes, No, Yes, Yes, Yes, No

5. At the end of the scanning, participants are allowed to guess object being drawn
6. The participant who has the most number of correct answers wins.

Possible Variations

More difficult drawings could be used.

Colour codes could also be used instead.

Practical Application

- What method/s did you use to draw the picture out?
- What ways worked, and why did the others fail?
- If someone gave you a lot of information, how should you listen to him/ her?

Game 42: Freezing Points

1. Players will sit in a circle. Have 3 different objects and assign them each a chemical.

CO_2 H_2O O_2

2. Each person is to pass the item representing the chemical with the highest melting point to the next person along the circle.

3. Pass the item with the 2nd highest melting point 2 people down; the item with the 3rd highest melting point 3 people down, etc.

Game 42: Freezing Points

Key Aptitude

Emotional Quotient – Co-ordination
Social Quotient – Teamwork, Communication

Math/Science Concept Applied

Chemistry – Freezing points

Equipment/Logistics

Balls, Pens, Straws

Time Required

20 minutes

Game Objective

Pass different items in a specific order while only one person is holding one object at any one time

Group Size

10-20

Procedure

1. Each object is labeled as 'the solid form' of different substances. For example, the ball is 'the solid form of water' (0°C), while pens and straws are 'the solid form of carbon dioxide' (–78°C) and oxygen (–219°C) respectively
2. The players are to sit in a circle

3. Each player will be given a ball each and they are told that they are holding solid water

4. When the game starts, the players are to pass the ball to the person on the right

5. All the objects should make three rounds at the fastest possible timing possible

6. As the game progresses, the players should pass the balls faster and the game master is to ensure that each person is holding only one item

7. To increase the difficulty of the game, the game master can replace a few balls with pens, telling the players that the pens are 'solid carbon dioxides' in Round Two

8. As CO_2 is colder, they have to pass the object to the second person on the right instead of the next person to the right

9. Once Round Two starts, the game master needs to ensure that each person is holding one object each

10. For Round Three, the game master can replace some balls and pens with straws, allowing three different objects to be used in the game

11. As O_2 has the lowest freezing point, it must be passed only to the third person on the right

Possible Variations

For younger players, the game master can simply name each object 1, 2 and 3. The object then needs to be passed to the next person on the right for 1, or second and third person on the right for 2 and 3 respectively

Practical Application

- How did you adapt to the changes during the game?
- What could have been done to improve the speed of getting things done?

Game 43: Mystery In The Food Web

1. One person sits out. He is known as the "detective".

2. Construct a food web and allocate an organism to everyone else, without others knowing what everyone else got. The end point of the web is the "human".

3. When the game commences, everyone will shake hands with each other and whisper to the other party his "identity". The prey between the 2 will sit out. If there is no direct relation, nothing happens.

Game 43: Mystery In The Food Web

Key Aptitude
Emotional Quotient – Flexibility
Social Quotient – Relationship Management, Communication

Math/Science Concept Applied
Biology – Ecology

Equipment/Logistics
Paper, Container

Time Required
20 minutes

Game Objective
To find out the identity of the players using the food web as clues

Group Size
10-20

Procedure
1. Choose a player to be the 'detective' of this game. He will have to find out who is the 'human' in this game is
2. Facilitator will have to construct a food web consisting of names of plants and animals living on the same terrain. The number of animals and plants in the food web must tally with the number of people participating

3. The food web ends with "human" as the top of the food chain. (e.g. a forest terrain consisting of apes, ants, elephant, ferns, deer, human)
4. On pieces of paper, write the names of the plants or animals in the food web constructed
5. Fold the pieces of paper and drop them into a container
6. Each participant then will pick one piece of the folded paper and the word bearing the name of a plant or animal is confidential to him alone
7. Each participant will then shake hands with another participant and whisper what animal or plant he represents to the other person and vice versa
8. If the animal/plant of the opponent is his 'predator', then the participant is "dead" and vice versa. If the animal of the opponent has no relationship with the participant's, then nothing will happen to the opponent
9. The game continues until there are 3 participants left
10. The detective has only a chance to find the "human" among the 3 of them
11. If the detective is correct, the "human" will be the next detective
12. If the detective is wrong, he has to continue to be the detective for the next game

Practical Application

- How did you feel when you had to drop out of the game?
- What did you learn from participating in the game?
- How could you have communicated more effectively?

Game 44: Static!

1. Each player gets a balloon and 5 pieces of paper. Rub the balloon against themselves to create static electricity and stick a piece of paper on the balloon.

2. They use their balloons to try and "attract" another player's piece of balloon.

3. After which, both players can "recharge" their balloons and repeat the process with others. When one loses all his pieces of paper, he has to leave the game area.

Game 44: Static!

Key Aptitude

Emotional Quotient – Goal Setting, Motivation
Social Quotient – Adaptability

Math/Science Concept Applied

Physics – Static electricity

Equipment/Logistics

Balloons, Paper

Time Required

10-20 minutes

Game Objective

To obtain as many pieces of paper as possible

Group Size

10-30

Procedure

1. Each player gets a balloon and 5 pieces of paper
2. The player will blow up his balloons and rub it against himself to create 'static electricity'
3. He will then stick 1 piece of paper on his balloon
4. The player will go around when the game begins to try and get another player's piece of paper off his balloon using only his 'charged-up'

balloon's static electricity to stick to his balloon. No hands or other means will be allowed

5. Once a player gains the opponent's piece of paper, he can remove the paper to re-charge his balloon again, repeating the Procedure 4 with another participant

6. If a player loses a piece of paper, he can recharge his balloon and stick another piece of paper onto her balloon and find another participant

7. Once a player loses all his pieces of papers, he has to leave the game arena

8. The winner is the player who obtains the most number of pieces of paper at the end of 10 minutes

Practical Application

- Mention two feelings you experienced when you lost your piece/pieces of paper. How were you affected?
- How would you strategize to be the winner the next time you play this game?

Game 45: Breathe!

1. Facilitator prepares a red cabbage pH indicator by soaking red cabbage leaves in water till it turns violet.

red cabbage

2. Place the pH indicator in 2 transparent pails, to be put at the end of the line. Players are given a straw each and will line up in 2 groups behind the start line.

xxxxx

xbxxx

O } pail with pH
O } indicator

3. Each player will have 30s to blow as much CO_2 into the pail as possible before the next player in the group goes. The group has to try to change the color of the indicator red in the shortest amount of time.

 huff

Game 45: Breathe!

Emotional Quotient – Goal Setting
Social Quotient – Teamwork, Relationship Management

Math/Science Concept Applied

Chemistry – Acids and bases

Equipment/Logistics

Red Cabbage, Water, Straws, Transparent Pails

Time Required

10-30 minutes

Game Objective

Turn the red cabbage water colour from purple to red

Group Size

10-30

Procedure

1. Prepare the red cabbage pH indicator by breaking up the cabbage leaves in water and letting it soak until the water turns a deep violet colour
2. Fill the transparent pails with equal amounts of the drained red cabbage water
3. Designate a start and finish line
4. Place one pail at the finish line of each group

5. Divide the players into teams of about 5-8 players each
6. Distribute a prepared straw to each player
7. Once the game commences, the first player of each team will run to the pail of red cabbage water and blow into the pail through the straw for 30 secs
8. After which, the game master will blow the whistle to indicate to the first player to return to his team-mates while the second player run towards the pail and start blowing for another 30 seconds
9. The team that manages to turn the water from 'purple' to 'red' first wins the game

Possible Variation

To increase the difficulty of the game, cut 2 holes in the straws near the top, so that when the players blow into the pail of cabbage water, some of the air escapes out of the holes

Practical Application

- As a team-mate, how did you support others in your team to achieve the goal?
- How did you manage your feelings when your team was losing?

Game 46: Flipside

1. Divide players into 2 groups. Cut open the garbage bag to achieve the largest surface area and have players hold the edges of the garbage bag.

2. A tennis ball will be placed in the middle of the bag. The group has to throw the tennis ball up, flip the bag, and catch the ball again.

Game 46: Flipside

Emotional Quotient – Co-operation
Social Quotient – Teamwork, Co-ordination

Math/Science Concept Applied
Physics – Gravity; Projectile Motion

Equipment/Logistics
Trash Bags, Tennis Balls

Time Required
10-20 minutes

Game Objective
To catch the ball with the trash bag

Group Size
10-20

Procedure
1. Divide the team into 2
2. Cut open the trash bag to open it to the largest surface area
3. Each player in the team holds the edges of the trash bag
4. The game master will place a tennis ball in the centre of the trash bag

5. The objective of the game is for the team to throw the tennis ball into the air and flip the trash bag to the other side before catching the ball with the trash bag again

Possible Variations

Ping pong balls can be used and more than one ball can be placed in the centre of the bag

To increase the difficulty of the game, the game master can place a tennis ball and a ping pong ball in the bag so that the team has to catch both balls after they flip the trash bag

Practical Application

- What were some of the strengths/weaknesses your team members showed?
- Are you expecting too much from others in the team to attain the goal? How did you manage your emotions?

Game 47: Same Train

1. Facilitator will prepare pieces of paper with "number terms" and distribute them to each player.

2. He will then call out number properties. Each player assigns himself a number of his choosing that fits his paper, and "combines" with 3 others to form a number that has the property mentioned by the facilitator.

Game 47: Same Train

Key Aptitude

Emotional Quotient – Communication
Intelligence Quotient – Problem-solving
Social Quotient – Co-ordination, Co-operation

Math/Science Concept Applied

Mathematics – Parity; Prime Numbers

Equipment/Logistics

Paper

Time Required

10-20 minutes

Game Objective

To form a chain as instructed as soon as possible

Group Size

More than 30

Procedure

1. Facilitator will prepare pieces of paper with number terms (e.g. Zero, Even number, Odd number, Prime number)
2. He will distribute the pieces of paper to the players
3. The pieces of paper denotes the number(s) available to the players

4. For example, when a player receives a piece of paper stating 'Even numbers', the numbers available to the player will be 2,4,6,8 during the game play
5. The facilitator makes statements about numbers when the players are ready (e.g. I can be divided by 25)
6. The players must form a chain of 4, with the first number in front, and the second number behind, putting both hands on the first player's shoulders, and so on (e.g. 3240 or prime number, even number, odd number, zero)
7. Once the train of 4 players is formed, the group has to shout 'Choo! Choo!' to get their number verified
8. Each person in a properly formed train will get a point
9. The player with the most points wins

Practical Application

- As a participant, how were you challenged when the question was changed each time?
- Could the communication between the participants be improved? In what ways could it be improved?

Game 48: Chemi-Who?

1. Divide everyone into 2 groups. Each group will send out a representative, who will be given an element card.

Element card

2. Each representative will turns to ask Yes/No questions to the other representative. Try to guess the element on the other card as quickly as possible!

Is it a noble gas?

No. Is yours a transition metal?

Game 48: Chemi-Who?

Key Aptitude

Emotional Quotient – Honesty, Persistence
Social Quotient – Social Awareness

Math/Science Concept Applied

Chemistry – Chemical formulas

Equipment/Logistics

Cards with the name of one of the elements in the periodic table on each of them

Time Required

30 minutes

Game Objective

The group has to guess the element names as fast as possible to earn a point. The group with the highest score wins the game.

Group Size

Minimum of 4 (preferably an even number)

Procedure

1. Divide the group into 2 teams
2. Every participant will pick a piece of paper that has the name of an element written on it from a box that contains all the elements from the

periodic table. The 'element' that the person has picked is only known to him

3. The first players from both teams take turns to tell the audience (their own team and the opponent's team) which element they are representing
4. The facilitator will decide if the information given by the players is correct
5. The aim of the players will be to guess the element in the fastest time possible
6. The two teams will first start by playing "Scissors, Paper, Stone" to determine who starts asking questions first. Questions asked must only require a nod or a shake of the head
7. The faster of the 2 teams to get the answer will win the round and gain 1 point for his team
8. If any information given by any player is found to be false, the opponent wins the round. The team with the highest score after everyone has played wins the game

Possible Variations

To increase the difficulty, instead of using elements, common compounds could be used instead (such as NaCl, also known as table salt)

Biology/ Physics: This game could also in the context of physics and biology replacing elements with physics formulas or animal diversity classifications respectively

Practical Application

- Will you expose your teammates if they answered the questions wrongly?
- Which is more important, integrity or success? Why?

Game 49: Limps In Motion

Everyone is squeezed on a mat. Facilitator will call out an object, and players must have the corresponding number of hands and legs.

A bee minus 1 leg!

Game 49: Limps In Motion

Key Aptitude

Emotional Quotient – Dexterity, Self-confidence
Social Quotient – Co-operation, Teamwork

Math/Science Concept Applied

Biology – Biodiversity of animals

Equipment/Logistics

Newspapers

Time Required

10-15 minutes

Game Objective

To achieve the right number of legs touching the ground for any given animal, insect or sea creatures

Group Size

3-4

Procedure

1. The players for each group will stand close to each other on 2 sheets of newspaper
2. When the facilitator shouts out the name of an/a animal/insect/sea creature, the players must put out the same number of hands or legs to touch the floor of the animal

3. No other parts of the body should touch the floor

Possible Variations

Instead of calling only single animals, game masters can request for 'two octopuses' or 'bee missing one leg'

The allocated area for the players to stand on could also be specified (such as on one piece of newspaper)

Practical Application

- What kind of difficulty or frustration did you encounter whilst trying to balance yourself?
- How did you help your teammates to get the task done?

Game 50: Melting Pot

1. Place signs of chemical symbols, plus and arrows face down in a circular area.

2. Divide players in groups of 2-4. The facilitator will name the product of a chemical reaction. Players will try to find the reactants of that product in the circle.

3. They will pass all the reactants to their group members to form a coherent reaction.

Game 50: Melting Pot

Key Aptitude

Emotional Quotient – Co-ordination, Respect
Intelligence Quotient – Memory Recall, Problem-solving
Social Quotient – Communication, Teamwork

Math/Science Concept Applied

Chemistry – Chemical Notation

Equipment/Logistics

Paper, Markers, Ropes

Time Required

20 minutes

Game Objective

To complete accurate formulas in the shortest time possible

Group Size

Any

Procedure

1. Facilitator will cordon off one large circular area with some markings on the floor, creating the 'pot'
2. In the centre of the play area, he will place cards with chemical symbols of different materials as well as cards showing 'plus' and 'arrow' signs

3. All cards should be placed facing down
4. Form teams with 2 to 4 players each
5. The teams will decide who enters the 'pot' and who will act as 'servers'
6. There must be an equal number of persons in each subgroup
7. The game starts when the facilitator names a product of a chemical reaction
8. For example, if the game master says carbon dioxide, players in the 'pot' can find the symbols for vinegar (CH_3COOH), baking soda ($NaHCO_3$), CO_2, H_2O, CH_3COONa, 3 plus (+) signs and an arrow
9. Alternatively, they can find FE2O3, CO, FE, CO2, 1 '2', 2 '3's, 2 plus signs and an arrow
10. If the symbol on the card the players has taken is not required, the player will need to place the symbol back
11. Once all the symbols and chemical formulas have been found, the players have to pass the whole set to the 'servers' who will form the complete formula to win a point
12. The game continues until no more complete formulas can be created

Possible Variations

To simplify the game, names instead of chemical symbols can be used

Mathematics: Mathematical symbols and numbers can be used instead if chemistry formulas. This game can be modified for younger players by using simple arithmetic

Practical Application

- How did you feel when your team won/lost?
- Could the communication between team members be improved? In what ways could it be improved?
- What were some factors that determined the success of the team?